HOPE

. . . even for Unitarians

HOPE

. . . even for Unitarians

At the turning-point of human evolution,
a rational moral compass,
a reason for being,
and a way to comfort

by David Sayre

*with an Introduction from the Conversations of
Werner Heisenberg*

Peter E. Randall Publisher
Portsmouth, New Hampshire
2021

ISBN: 978-1-937721-74-9
Library of Congress Control Number: 2020924859

Printed in the United States of America

Published by
Peter E. Randall Publisher
Portsmouth, New Hampshire 03801
www.perpublisher.com

Book design by Grace Peirce

To Polaris

The Winter Solstice
Vermont, 2020

Contents

Introduction

We Unitarians are full of uncertainty.

We'll take our introduction, then, from the scientist who knew more about uncertainty than anyone. Werner Heisenberg is best known for the Uncertainty Principle that bears his name. He won a Nobel Prize "for the creation of quantum mechanics" and wrote this about our moral compass:

> The problem of values...concerns the compass by which we must steer our ship if we are to set a true course through life...I have the clear impression that all such formulations [of philosophies] try to express [our] relatedness to a central order...And when people search for values, they are probably searching for the kind of actions that are in harmony with the central order... It is in this context that my idea of truth impinges on the reality of religious experience. I feel that this link has become much

more obvious since we have understood quantum theory.[1]

What shall we make of Heisenberg's instinct for a "harmony with the central order"? He was no mystic, and he did not intend this as metaphor. Nor was he alone: Einstein, Planck, Bohr, Schrödinger, Pauli, and others shared a sense of some kind of "central order," as Heisenberg described in his *Conversations*. They had a deep understanding of order in the universe. They saw some connections that few others understood.

Five decades later, we can begin to piece together what they saw in a new light. We have lived with the strange realities of relativity and quantum physics for several generations now. We can test their theories, and we have put them to good (or not so good) use. They were right.

---

1. Werner Heisenberg, *Physics and Beyond: Encounters and Conversations* (Harper & Row, 1971), 214–216. Heisenberg's work in Germany after Hitler's ascension has long been a matter of debate, ranging from praise to condemnation. A balanced perspective can be found in Michael Eckert, "Werner Heisenberg: Controversial Scientist," *Physics World* (November 30, 2001). https://physicsworld.com/a/werner-heisenberg-controversial-scientist.

Heisenberg and his quantum colleagues might have filled in the logic of harmony with a central order, but they ran out of time. The world took their theories and made nuclear energy and quantum computers and all manner of modern things that everyone uses and no one really understands.

And now we too are running out of time. We can't see inside all our artificial-intelligence programs any more. The world's climate is unstable. The power of destruction is held by a few leaders of limited wisdom. The human genome lies on our lab tables, open to splicing for good or ill.

Our grandchildren stand at this hinge of human evolution, now for the first time in their hands. What shall we say to them?

*When people search
for values, they are
probably searching for
the kind of actions that
are in harmony with
the central order.*

WERNER HEISENBERG

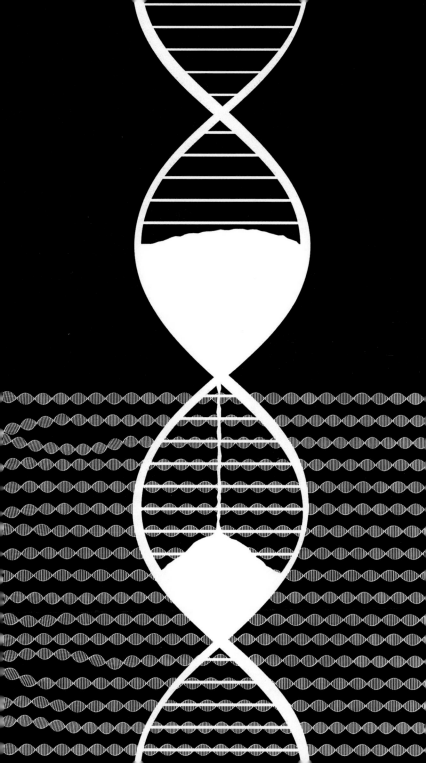

Looking for Intelligent Life in the Universe

In 1985, producer-director Steven Spielberg had a problem. How would he promote a new movie about "The Extraterrestrial"? The title was too long. It was hard as well to follow the big success of Kubrick's *2001: A Space Odyssey*. He needed a link to real science.

His solution included a $100,000 donation to Carl Sagan's Planetary Society to advance the "Search for Extraterrestrial Intelligence" project; a trip to Harvard, MA, with his infant son Max to dedicate the SETI radio telescope with Sagan; and naming his movie *ET*.[2]

I visited the SETI telescope a few years later and wrote a chapter about it,[3] but our search took a wrong turn. The public will pay to see a movie

---

2. Ann Levison, *Harvard Post*, October 4, 1985.

3. David Sayre, "Harvard, Massachusetts," in *Something There Is*, 2nd ed. (Peter E. Randall Publisher, 2014), 7–12.

like *ET* and to look for watery planets that might harbor "life" that looks like us (or even like ET). But that's not what Carl's radio telescope looked for. He knew (as did Heisenberg) that *intelligence* in the universe would not be limited to forms like ours, or brains like ours, or even to individual packages. That's a Flatland perspective.[4]

The SETI radio telescope at Harvard, MA

Carl Sagan was familiar with Heisenberg's work, of course, and was skilled himself at movie and TV scripts. He could easily have imagined more astounding stories based on

4. See the illustrated children's book *Flatland* (Two Little Birds Books, 2014) by the wonderful artist Rebecca Emberley and me (recommended by Parents' Choice Awards). Also, our companion work, *The Flatland Dialogues* (Peter E. Randall Publisher, 2017).

weird quantum phenomena. Particles coming in and out of existence, for example, or the entanglement that Einstein called a "spooky action at a distance," or Schrödinger's famous cat that might be both alive and dead at the same time. But Spielberg had a keener sense of public taste, and his investment in *ET* paid off handsomely.

The SETI project continues, but, so far, there are no reports of intelligence beyond Earth. What does that tell us? (1) There's nobody out there, and Earth-brains are the only intelligence in the universe; or (2) Others don't do anything we can yet detect. Since (1) seems very unlikely, we have to consider other possible forms of intelligence. We have to look farther.

Indeed, other expressions of intelligence may be all around us, but we can't detect each other. It's unlikely that others send our kind of radio signals, use our kind of communication media, or leave other evidence we can understand. It's unlikely they use the same senses that evolved to survive here on Earth. So, although it may be interesting to find biological traces on other planets, that will not tell us much about intelligence itself.

Our understanding of intelligence—of reality itself—was greatly expanded by Heisenberg and

Other expressions of intelligence may be all around us...

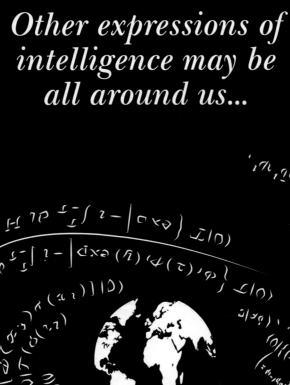

but we can't detect each other.

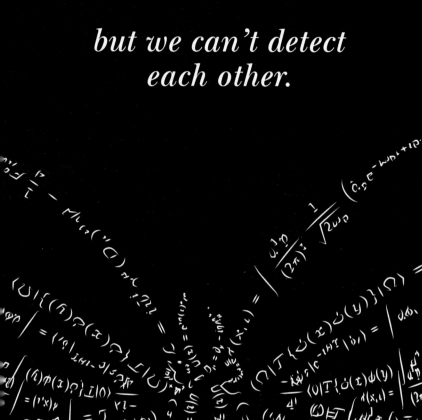

his pioneering friends. Their discoveries showed things that still seem bizarre to us—particles can be "entangled" even when separated, electrons can tunnel through barriers, light can be two things at once until detected, we can't know things precisely, etc. This prompted all sorts of explanations, struggling to fit our experience. Maybe human observation changes what happens. Maybe the world branches when we measure a quantum particle, going off in different futures to fit the outcome. Maybe (as Einstein thought) there are hidden variables that will make the behavior understandable on our terms. At the time, the quantum theories could be verified and (perhaps) understood only through their math.[5]

Over the years, we've confirmed the theories in tests and used them in products. And we've

---

5. We don't normally see the behavior of quantum-size particles because they bump into too many other things—and we're a collection of a zillion other things. But their ability to be "entangled" with each other, their "superposition" of more than one condition, etc. are real, and we are beginning to use those realities in things like quantum computers and communications. A clear, de-mystifying (but authentic) article on this realization—in a non-technical journal, without any equations—is offered by Philip Ball, "The Universe is Always Looking," *Atlantic*, October 2018.

come to two better realizations:

1) The quantum world is real. It just seems weird because we don't recognize its effects on us. In fact, we don't recognize a lot of reality. Gradually, we have realized that our day-to-day sense of place and of time is *incomplete*.

2) That should not be surprising. Our experience comes to us from senses and brains that evolved by "natural selection" to survive on Earth. That's all we needed. But now we know there's far more to reality than we see, even with our advanced instruments.

It would be silly to believe that we can sniff out all that's real. Or feel it, or taste it, or hear it singing to us, or see all the expressions of intelligent life in the universe. Or even theorize about the whole of reality with our evolved brains (or ones we fabricate). Like the characters in our *Flatland* book, we can't see off our page.

But we know for sure there's a lot above our page, and it's real. We have to look farther.

Is there a way to look up? To stretch our experience? To see beyond what Shakespeare called "this muddy vesture of decay"? Let's look hard at that decay.

We can't see off our page,
but we know for sure
that there's a lot
above our page,
and it's real.

The Bug

Long before Heisenberg and quantum physics, scientists and philosophers and Unitarians knew about a peculiar bug in our universe. This bug is all about decay. It doesn't like order. It eats away at order everywhere. And it keeps growing. You can shrink it in one spot, but that makes it grow in other spots.

Some of us spend our whole careers struggling with this bug. It's what makes losses in energy, and noise in communication. We can't get rid of it. We can move it around, however, and we've learned a lot recently about it. How to measure it, for example. And most important, what it has to do with intelligent life.

Of course, I'm talking about [shh—Entropy]. But if I write its name or its formulas, folks will stop reading. So, I'll call it the Bug.

If there were genders in physics, the Bug would identify as they/their/them. The Bug assumes three identities. Yes, and we see them in three costumes: disorder, ignorance, and time.

11

*This bug is
all about decay.*

It doesn't like order.

For centuries, only the first costume was noticed. We all recognize it. Whenever we change anything, we can't put it back exactly without losing something. It's easy to pick a blossom, almost impossible to put it back. If I store energy somewhere—say in a high pond, or a battery—I always get back less than I put there. (I do a lot of that.)[6] We could call that first costume that the Bug wears "disintegration," or decay, or just *disorder*.

This tendency toward equilibrium, toward disintegration of all structure—toward death—is universal. There is nothing magic or even mysterious about this "second law" of thermodynamics;[7] it is simply a natural trend toward the most probable state of all things. Structure is improbable, chaos probable. Sharp distinctions are improbable, equilibrium probable. In

6. You might enjoy my description of the Troll under Northfield Mountain in *Something There Is*, p. 237. The following paragraphs are borrowed from that description.

7. Science students are taught early four fundamental laws of thermodynamics. The first law just says we can't create or destroy energy (in an isolated system), but can only change its form. The second law says the disorder in such a system can only increase. The third and fourth laws deal with defining temperature and what happens near absolute zero.

each case we have to expend far more energy to sustain order than to allow disorder. There are simply more ways to get to a disorganized condition than to keep an organized one, so its probability is higher—very much higher, as it turns out.

Whenever any system or mixture "settles down"—reaches equilibrium—we find that distinctions have blurred, differences smoothed out. A glass into which yellow and blue fluids are poured from opposite sides quickly turns green, and stays that way, because there are enormously more "mixed up" arrangements of molecules than the few that will still look yellow and blue. All the molecules keep moving around and bumping into each other, under the influence of the Bug (chaotic heat-energy). The probability that the yellows and the blues will ever again be found arrayed on opposite sides is extremely low. (In just a thimble of a simple gas at ordinary temperature and pressure, the number of such molecular collisions is on the order of 1,000,000,000,000,000,000,000,000,000, 000,000 every second, which gives some idea of the low probabilities associated with maintaining distinct, segregated arrangements.) We could force the yellow and blue molecules back to their original places with a sufficient mechanism for detecting and catching and moving them, but it would require hugely more directed energy than the random heat energy that keeps them mixed up.

When a clock's pendulum swings, it cannot regain its initial height without our winding a spring to re-inject the energy it lost to frictional heat. When I run downhill, I never recover all the energy that I expended climbing the hill. The Second Law says that all clocks will eventually stop, all blossoms decay, all runners tire, all structure break down to its lowest state.

So, we all know that we lose something in every transaction. But how much? One hundred and fifty years ago, Ludwig Boltzmann proposed a solution to that question, and a form of his answer is carved on his gravestone.[8] It has been extremely useful ever since, to anyone designing or describing energy-using systems. It tells us how much we must lose in any process, how much useful energy we simply can't get out of it—the size of the Bug. Roughly speaking, it's proportional to the probabilities of all the conditions that all the little pieces of anything (like the molecules in that thimble of gas) can take. It's a measure of disorder. Things that are

---

8. $S = k. \log W$

perfectly arranged, frozen in place, score low; when we heat them up, jumbling all their pieces around randomly, their score goes up.

There is no escaping the Bug, as Sir Arthur Eddington eloquently commented ninety years ago:

> If someone points out to you that your pet theory of the universe is in disagreement with Maxwell's equations—then so much the worse for Maxwell's equations. If it is found to be contradicted by observation—well, these experimentalists do bungle things sometimes. But if your theory is found to be against the second law of thermodynamics I can give you no hope; there is nothing for it but to collapse in deepest humiliation.[9]

We're not dealing with metaphors here. These are hard facts. I spent a couple of pages on them, because they're the reason we have

---

9. Arthur S. Eddington, *The Nature of the Physical World* (Cambridge University Press, 1928), 74. The famous "Second Law of Thermodynamics" basically says that the total "entropy" (disorder) of any isolated system can never decrease, can remain constant only if everything in it is entirely reversible, and therefore inevitably increases. Every isolated system, from a cup of tea to the universe, evolves spontaneously toward "equilibrium," the least orderly state.

death. And the Bug doesn't stop there. It has two other gifts for us, in its two other costumes: ignorance, and time.

More recently, we've noticed the Bug's second costume: ignorance. More accurately, it's the information we don't know in advance—the information contained in any message (or painting, symphony, or anything else we can learn from) before we figure it out.[10] We could call that costume "unknown information," or just *ignorance.* And we can measure that, too.

In 1948, Claude Shannon was wrestling with what seemed a very different problem from Boltzmann's: how fast one could push a message through a cable or other limited channel. But once into it, he found an elegant way to measure how much new information is contained in anything we're trying to understand. Scientists now refer to this as the "Shannon entropy."[11] It

---

10. I offer a reasonably accessible account of this in the "Radio Communications Corporation" chapter of *Something There Is*, pp. 28–40.

11. Shannon's "A Mathematical Theory of Communications," is generally considered the foundation of modern information theory. He consulted the great statistical thermodynamics expert John von Neumann about what he (Shannon) should call his new measure of information content: "I thought of calling

applies to everything. Stephen Hawking even used it to compute the information contained in a black hole.

The information content in a message (or anything else) is what we don't know about it. In a word, our ignorance of it. Once we've received and understood the message, its information content becomes zero. We exchange information only when there is some surprise, some uncertainty, in what we are about to hear. Information is entirely missing from messages that are known in advance, and at a maximum when they're least predictable. With increasing familiarity, anything communicates less. A joke tires very quickly because its surprise wears out: I've heard that one before.

Like Boltzmann calculating unavailable energy, Shannon figured out how to calculate

it 'information', but the word was overly used, so I decided to call it 'uncertainty'....Von Neumann told me, 'You should call it entropy, for two reasons. In the first place your uncertainty function has been used in statistical mechanics under that name, so it already has a name. In the second place, and more important, nobody knows what entropy really is, so in a debate you will always have the advantage.'" (Quoted in *Scientific American* by M. Tribus of MIT in 1971 ["Energy and Information"], Vol. 224, 1971, pp. 179–188.)

unknown information. Both are proportional to the probabilities of the conditions of all the little bits that make them up. Shannon calculated that information content, and it's been extremely useful ever since.[12]

Now here's a really cool thing. The measure of disorder in any isolated thing comes out exactly equal to Shannon's measure of information! Disorder and ignorance are equivalent. In fact, disorder can be considered a special case of ignorance.[13]

Why are we excited about this? Because we're looking for intelligence in the universe, and for how that's related to a "central order." Ignorance and disorder, get it? Intelligence and a central order in the universe. It would be logical to consider how those are related. And what their relation might reveal—what may not be infected by the Bug, after all. We have to look farther.

---

12. For those interested, Shannon's formula looks like this: $H(X) = -\sum_{i=1}^{n} P(x_i) \log P(x_i)$

13. I can't prove this in a sentence or two, but there's vast literature about it. Among others, Edwin T. Jaynes in 1957 suggested that thermodynamic entropy be considered an *application* of information theory, ("Information Theory and Statistical Mechanics," *The Physical Review,* May 15, 1957.)

It took me a long time to get my head around this. A couple of careers, luckily in communications and energy. (Both those sciences are all about displacing the Bug.) A lot of really smart colleagues. The loss of loved ones, and a recovery of hope.

In fact, disorder can be considered a special case of ignorance.

What Is Intelligent Life, Anyway?

OK, let's slip into that little shack behind Sagan's radio telescope, almost hidden in the trees. Turn on all the computer screens. Strap on the headset. Listen to all the signals from the sky. We have an important job: to send out an alarm when we detect evidence of intelligence somewhere.

Is it likely such "extraterrestrial intelligence" really exists? Or let me put it differently: Is it likely that our brains are the only expressions of intelligence in the universe? Probably not, you say. Then what would convince us? What evidences of intelligence would the SETI tele-scopes consider real, as they peer around our universe? We can't see a smile or a wave, or sniff or hear or taste or touch intelligence out there. We're searching for intelligence in the universe, not for some packages that seem to display it.

How will we recognize it? Hmm. Well, as a start we'd want to detect some sense of order, right? We'd look for an ability to organize, to have some purpose, to affect outcomes. To distinguish between random chaos and some structure and to cause the latter. Probably many "life forms" exist out there with such a capacity in this very large universe, don't you think? They don't have to be carbon-based, or water-drinking, or even physical as we recognize physique.

Let's look for some.

Wow, we can see lots of beautiful crystals out there, and regular beep-beep-beep, and constant humming, and...OK, order by itself is not enough. The universe is full of orderly crystals and large-scale structures that do not answer our questions.

How about "coherence"? We nerds love coherence—predictability, sticking together, consistency...but no, that's not enough either: the radio telescope at Harvard listens all day to perfect monotones in the cosmos, carrying no information at all.

What about lots of new information? Nope. The more random and noisy a signal, the more information it theoretically carries, and there's sure a lot of noise—but no purpose to it.

Gradually we realize that we need evidence of four things together:

1) an ability to *organize*, to recognize and bring order out of random, noisy, chaotic processes;

2) *freedom* to choose such order;

3) an ability to *communicate* it in some form; and

4) an intention to do so, an indication of *purpose*. Then we'd be convinced that we've detected an intelligent process out there somewhere.

You may have noticed that these all deal with reducing the Bug's bite! When we look for intelligence, we're looking for an intentional reduction (or at least displacement) of disorder and ignorance. For evidence of increasing order and decoding information.

It's a signal carrying that kind of evidence that the antenna on Harvard's hill patiently waits for. Organizing, free, communicating, intentional signals. Each is necessary, but not sufficient in itself. (By contrast, being a human on Planet Earth is sufficient, but not necessary, right?)

When we look for intelligence, we're looking

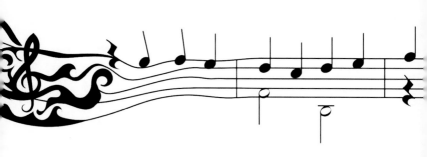

for an intentional reduction ...of disorder and ignorance.

No doubt other such products of evolution—not to mention the brilliant "robots" we can increasingly build—are expressing intelligence around our universe and others. Each probably considers itself unique. Each likely has a model of the universe that it discovered and considers complete. There is not even a bright line between "artificial" intelligence and what we might like to consider the real thing. There is nothing sacred about the evolution of human brains from the mud of earth, or any other form of mud anywhere.

I gave a talk about this on Star Island a couple of summers ago at an Institute on Religion in an Age of Science (IRAS) conference on "artificial intelligence."[14] Intelligence is what lets us put things together out of chaos and detect messages out of noise. It pushes away the Bug in both costumes: disorder and ignorance. It would be possible only in a rational universe, based on invariant laws—a central order.

That central order is there. We didn't put it there. There is something, rather than nothing. That something is not capricious, not arbitrary,

---

14. David Sayre, "Another Way to Look at IT," (lecture, Institute on Religion in an Age of Science – Artificial Intelligence Turns Deep, Star Island, NH, June 2018.)

not magic. It is orderly, understandable, logical. It was there before we were. And it's not limited (as we are, in the thin slice of reality that we experience). There's a reality of universal laws that we can count on. Intentional, intelligent processes can always build order.[15] Heisenberg and his colleagues realized that intelligent life, whatever form it might take, must be an expression of that kind of central order in the universe.[16]

Heisenberg was an accomplished musician as well as a scientist. When he spoke of being "in harmony with the central order," he had in mind the way we respond to a well-tuned orchestra, or an insistent beat. They strike resonances in our hearing and our experience. We want to dance, or sing along, or just engage with it emotionally. We can even join in. If we have the skill, our joining is harmonious—it adds in a way that scientists call "coherent." Or perhaps

---

15. OK, for you nerds I'll add "at the expense of increasing the surrounding disorder."

16. By "expression," we mean the way genes are expressed in our cells, the way an underlying truth or principle is manifested in practice, the way a potential is realized. The central order and truth are present in every event but are not always recognized or demonstrated or acted upon—not always *expressed*.

That central order is there. We didn't put it there...

It was there before we were.

we are off-key, or off-beat, in which case we interfere instead of adding.

That kind of coherent combining or integrating—adding vs. interfering—shows up in all physical systems, out in the cosmos and down among quantum states of matter, and in all we do. We'll come back to it.

Anyway, the sound of music sets up an acoustic "field" in the air around us. Things in that field (including our eardrums) start vibrating at its frequencies. Heisenberg realized that a central order would set up that same kind of potential everywhere in the universe—a "field" of intelligence.[17] Any expression of intelligence, including us, might be more or less attuned to that field—in or out of "harmony" with it, in his words.

Everywhere we look as scientists—among us, beyond us, within us—there are these laws, this central order, this making sense. We have not found any time or place without them. We don't expect to find any spot that's senseless, without order. If we could travel to a distant planet, or a different time, we could look back and observe

_____

17. To physicists, a field is something with a value and/ or direction at every point in time and space.

the same laws, the same effects, the same order, at every point in time and space. Any intelligent observer, including those that the SETI project might find, could agree on their effect.

It takes intelligence to recognize or bring about order, and we find evidence of a central order at every point.

Does the universal potential of intelligence not seem like a real field? Well, we've learned a lot about fields lately, and the idea no longer feels remote or unfamiliar. We now carry phones that work when we're within the field set up by our carrier's cell antennas. Same with the earth's gravitational field, which holds us down.[18] Food in a microwave oven heats up when its water and fat molecules vibrate at the oven's field frequencies (fast). Recently, we've detected the particle associated with the "Higgs field."[19] Without that field, we'd all fly apart aimlessly at

---

18. We've recently detected gravitational "waves" that Einstein predicted would be launched from massive collisions in the cosmos.

19. Leon Lederman, head of the Fermi Lab, reluctantly called this "the God particle" in his delightful book of that name with Dick Teresi (Dell Publishing, 1993). The Higgs effect was theorized back in 1964, but its particle was finally detected only in 2012.

the speed of light.[20] All these "fields" are part of our lives, even though we can't see them.[21]

This making sense, this universal logic, this order in things, makes everything possible. We can organize and learn anywhere. We can communicate and reach out to each other, because we're in touch with the same universal truth, the same intelligence-field. That reality has no boundaries, it's not prevented anywhere, at any time. We could call that field of intelligence-potential, as Heisenberg did, a "central order." All intelligent life depends on it.

I'm not a theologian. My point is not to propose (or oppose) some kind of God. It is simply to follow a trail of logic that is hopeful as well as rational. That logic has been confirmed for me through long years of practice in entropy-reducing technologies.[22]

---

20. All forms of matter slow down—gain mass—as they struggle through the Higgs field, everywhere in our universe.

21. There are lots of fields and sub-fields that we know about, and probably some we don't (given the large reality of "dark matter" and "dark energy" that we can't yet explain. At least I can't.)

22. That practice—and the insights of many recognized experts—are recorded in *Something There Is*.

We can now be confident that there is far more to reality than we experience. Our day-to-day sense of time and place, and all that may fill it, is incomplete. We can also be confident that the freedom to choose order and communicate with some purpose—which we call intelligence—is everywhere available. It's a reality in itself. It's not just the sum of our limited brains.

That reality was there before our brains evolved, and it will still be there even if we blow up our planet and all intelligent life on it.

So that leads us back to Carl, Steven, and baby Max, in Harvard, MA. They were dedicating a radio telescope to look for expressions of that intelligence-field. Their hope was to detect some kind of evidence of intelligent life beyond Earth. It would not have to look like us, but it would have to show that it can deal with the Bug.

We can communicate ...

because we're in touch with the same universal truth.

Capacities of Intelligent Life

Let's say that one of the SETI scientists happens to come into Sagan's processing shack while we're still in there.[23]

Skeptical SETI Scientist: Hi. I hear you're interested in how we'd detect intelligence.

Us: Well, we've figured out how it would be expressed, if you can detect it. The recognition of order, the freedom to choose it, the ability to communicate with some purpose, at least. What would you say they have in common?

SS: That's easy. They all reduce what we call "entropy." We experience entropy in two forms, you know—the disintegration of everything unless you build it up, and the information content in any message until you receive and decode it (not to mention time itself). Decay and ignorance, if you will. They're mathematically

23. The following dialog is adapted from *The Flatland Dialogues* by David Sayre and Rebecca Emberley (Peter E. Randall Publisher, 2017).

equivalent, you know. You can reduce entropy in a given system at the expense of a greater increase in the surroundings...well anyway, intelligence is what lets us put things together out of chaos, and detect messages out of noise. That's what we're looking for.

Us: And where would you say these things might come from?

SS: I would say from the physical and mathematical laws that are invariant around the universe. We believe in a rational universe based on those laws. The universe is probabilistic, but not capricious. By that, I mean there's a reason for everything that happens, even though we may not be able to see it. We don't believe in miracles...I hope you're not going to tell me that there are mystical wonders or things we have to take "on faith"—?

Us: No, no. It's that rational universe that supports the things we care about. A reality that exists of itself. But what other experiences would reduce entropy among intelligent beings?

SS: Well, given the freedom to choose order and communicate, organizing and building are possible, even likely. And that leads to learning and teaching...and I'd say eventually healing each other, and so forth. If an intelligent society or association of some sort is free to organize, learn, communicate, and heal each other, they could keep reducing entropy locally.

Us: Do you think there are any out there?

SS: We think there are a lot of them that we haven't yet found. Some a lot more advanced than we are. I have no idea what form they might take, and I don't really care. Could be just an orderly cloud of energetic particles,[24] or some other way to organize, don't you think?

Us: Yes, of course. And to survive, it seems to us that they'd have to develop a farther reach in each other—do you know what we mean?

SS: I know what you're getting at. Things like love and beauty are important to us scientists, too. But we have to explain them rationally, in the context of reducing entropy. So, take love, for example. It can be seen as a combination of commitment and trust and a kind of "extension" to each other. By that, I mean what the psychologists call "identifying"—empathy, sharing experience, feeling like part of each

---

24. The astrophysicist Fred Hoyle envisioned that way of expressing intelligence in his novel *The Black Cloud* (William Heinemann, Ltd., 1957).

other. The physics metaphor would be what we call "entanglement"—two or more particles that can be described only as a single system, and that behave that way.

Us: So you think that love is real, and not just on Earth?

SS: Sure. In our experience on Earth, it's most intense between two persons, but any society needs that kind of stretch to each other, to reduce disintegration and ignorance, to survive. Logically, love should add to each other. It should make us "more." By that I mean it should enhance each other's expression of the qualities of intelligent life. It should stretch our experience, our reach, our effect. Each member then expresses intelligence more cooperatively, so together we become more than we were alone. So, I'll grant that love is real in that sense, and essential...Or did you have some more spiritual explanation in mind?

Us: No, we like your explanation. Love is an expression of intelligence. To be real, it must be rational, which to us means faithful to truth. And what about beauty—that may be the hardest thing to explain, don't you think?

SS: I've thought about that, too. I know many

folks like to think beauty is beyond explanation, is mystical, some kind of spiritual experience that we shouldn't try to rationalize. But to me, seeking to understand anything is the greatest respect we can pay it. If beauty is real, if we hope to find it among all intelligent beings anywhere, we have to be able to say what it is, to communicate about it. Would you agree?

Us: Actually, yes. "Beauty" should not be based on some set of rules or standards, right? Nor should it be considered beyond our ability to understand. And like love, it would not be contained in what seems its source. Yet it's not a law of itself. So, it must be a capacity of intelligence. You'd expect to find it wherever you find intelligence expressed, right?

SS: Yes, it's almost like an instinct to us. Over and over, discoveries have come from an expectation that an "elegant" solution should exist. By that we mean an understanding, often in mathematical terms, that captures a lot of information in a simple expression. Think of Einstein's famous formula for the equivalence of energy and mass, or Claude Shannon's formula for the information content in a message—just brief little formulas that say so much! And when we find it, there's a thrill, a sense of wonder. The

Love is an
expression of
intelligence.

astrophysicist Chandrasekhar is often quoted as "shuddering before the beautiful...[that] a discovery motivated by a search after the beautiful in mathematics should find its exact replica in Nature..." This seems possible only in an orderly universe, only where we all have a direct route to the same self-existing truth, as Roger Penrose likes to say. Are you with me?

Us: We are. But mathematics and nature are not the only routes...

SS: I know, I was getting to that. We scientists often have a special reverence for the arts, for the same reasons. They evoke patterns or meaning in what would otherwise be chaotic. And at the same time, they communicate intensely about that meaning. In other words, they reduce both forms of entropy! Do you know why I consider that so important?

Us: Maybe because that evokes all the capacities of intelligent life? The freedom to organize, to learn and communicate truth, to heal and to love, all involve a choice of order and relation. They all reduce disintegration or ignorance— the two ways we experience entropy. So your trust in beauty reflects your faith in an orderly universe, in the shared experience of being in touch with a single truth. Is that what you had in mind?

SS: Yes. So, to me, beauty is a startling connection with each other, with the whole of which we're part, as Einstein said.[25] It should be a way to express intelligence more fully, to experience what's true, to share it. Therefore, beauty is rational, not mystical. Well, I get a lot of arguments about that, but how do you feel?

Us: We agree you've found a truthful and respectful way to think about love and beauty, along with the other expressions of intelligence. The important thing is that they are real. They can be understood. They are universal. They're not things invented by humans. So, as expressions of intelligence, they can be found anywhere, would you agree?

SS: We would expect those to be universal capabilities. We would expect to find them in other intelligent life-forms. That doesn't mean they exist on their own, like the laws and order of the universe. They are functions of intelligence; they need intelligence to be expressed. I hope you're not going to tell me that beauty and

25. This Einstein reference is quoted early in the chapter "Hope," below.

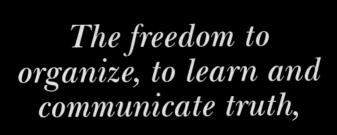

The freedom to organize, to learn and communicate truth,

to heal and to love, all
involve a choice
of order and relation.

love and freedom and so forth are some kind of Platonic reality that exist independent of intelligent beings—?

Us: No, if our people are to hold something sacred, it should be intelligent life itself, and truth. It should be our capacity to experience those things, to choose them. That freedom, that seeing and making and giving, are realities of intelligent life. They exist in our expression of intelligence, of truth, of what we might call a central order.

We think the sacred thing about beauty is that we can see it, and even make it. That is, we can all connect to that same truth, we can all express that same intelligence-potential. They are freely available to us.

And through them, because of them, we can *connect with each other*. By "each other," we mean all expressions of intelligence. The "whole each other," in the Einstein sense. All who reach for the same self-existing truth, the same central order.

That connection, that reach, that inclusion, may be all the "sacred" we can hope for. But perhaps it's all we need.

Moving the Bug in Our Neighborhoods: The Microgrid Lesson

You are alone and immobile. Your room suddenly goes dark, the sound of ventilation stops. Cell phones don't work. Oxygen runs out, medical devices fail. Temperatures fall fast—or rise fast, if in summer. You are trapped. The power grid has failed, and many will die.

This is not fantasy and not some distant country. We all risk losing both energy and communications as our planet's climate grows more unstable. The Bug's full fury is on display, tearing things down, spreading confusion. The risks are especially faced in low-income neighborhoods, by immobile residents.

As alleged experts in both technologies, my colleagues and I should be able to help. Right?

Well, not so easy.

We can fix the wires and stuff. We can raise the capital. We've worked in lots of scary and powerful places—prisons and mental institutions, cathedrals and laboratories, neighborhoods of the very rich and the very poor. We thought we understood them all pretty well. So, making folks' power stay on and their cell phones keep working should be easy.

It isn't. And it's not just because of the bureaucracies in utilities and governments. It's that Bug again, in yet another form.

So: Boston's Chinatown, and Chelsea, Massachusetts, are largely immigrant communities, and both are at great risk. Increased flooding in storms, heat waves, loss of power and of communication threaten immobile residents and the small businesses and institutions that serve them. They have seen the devastation of storms in New York and Puerto Rico, where they have deep connections. They decided to take charge.[26] They would let us help.

Maybe we could develop neighborhood-run "microgrids." These could make power and phones more reliable.

---

26. See Max Jungreis, "Chelsea residents see a powerful lesson in Puerto Rico's 2017 hurricane devastation," *Boston Globe*, October 14, 2019, https://www.bostonglobe.com/business/2019/10/13/memories-maria-power-plans-chelsea/a1RTk6uzNJHAk-KNEzRJP1K/story.html.

In fact, there are many microgrids in the world, more than one claiming to be the first.[27] But few are designed and run by the neighborhoods themselves. Their fix doesn't reach everyone, and it doesn't last. Why?

Because this isn't really about microgrids, or even about power or cell phones. It's about something harder. We call it "coherent integration." (Forgive the nerdy term. Just think of it as the opposite of disintegration.)

It turns out that there are organizations in these neighborhoods who have already designed a more effective "microgrid" without calling it that. What makes it work is this: they're reducing the Bug's bite out of two things at once.

The first bite we're familiar with: conventional power and phone systems wear out, they're subject to storms and mischief, they

---

27. A microgrid is just that—a relatively small group of energy consumers, usually interconnected on a campus or a street or some other clustering. These consumers can be interconnected and "islanded" from the main electrical grid. They can often use their own power sources, and can disconnect as a group to operate independently in case of a grid failure. (Our design is a "microgrid without borders," not requiring wired connections.)

*It's about
something harder.*

*We call it
coherent integration.*

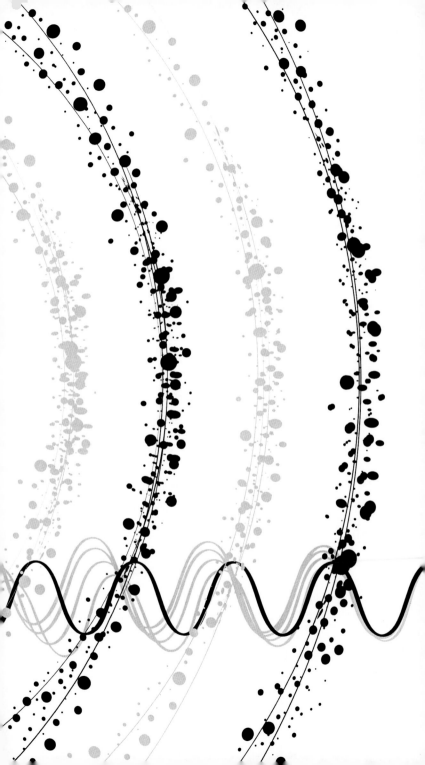

lose a lot of their stuff in transmission. All Bug-bites, which the neighborhoods can treat. They're making energy and communications local and democratic. We understand that, and we can help.

The second bite we hadn't thought of. The Bug isolates and confuses human efforts. It takes work to organize folks in a common purpose. It takes work to sustain all members' commitment to each other. The Bug is all anti-work.[28] It knows that everything falls back to its lowest-energy state and isolation if left alone. But all the things we've been talking about—building useful things, communicating, choosing freely, sharing a common purpose, healing, trusting and committing—all these things reduce the Bug's bite. And it seems they're what these neighborhood organizations are all about.

This common commitment to each other's expression of intelligence isn't just efficient. Inclusion of diversity isn't just a slogan.

---

28. Engineers measure "energy" and "work" in the same units, but use the term "work" to mean *useful* energy. I expend a lot of energy running up a hill, but only the height gain is useful work—the rest goes into wasted heat. That disorganized heat, divided by temperature, is a rough definition of "entropy" (our Bug).

It displaces the Bug. It's in harmony with the central order.

Based on that commitment, the neighborhoods adopted a new design for distributed microgrids. This is why we're excited about it, why we've dropped other things to focus on it. The neighborhoods' principles include these:

Any home or small business or institution should be able to join, not just those that could be wired together. The whole building and those in it should be protected, not just "emergency circuits." Participants' costs should be reduced first, by increasing energy efficiency. Fossil fuel burning should be eliminated. Low-income participants should not be charged more than they save. Utility wiring and utility control over microgrid operations should not be required unless they agree to buy the benefits to their grid. There should be back-up cell phone service and charging in any emergency. The community should own electric vehicles and chargers. Neighborhood residents should be trained and employed in clean-energy installations and service.

This ain't your father's microgrid. The need goes way past what he had in mind. Energy and communication are matters of life and death

Building useful things, communicating, choosing freely, sharing a common purpose, healing, trusting and committing—

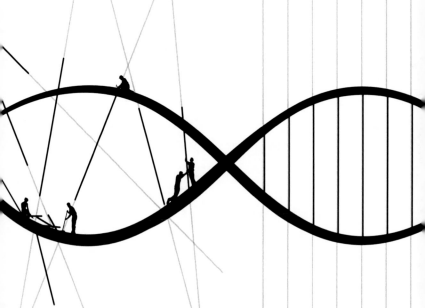

all these things reduce the Bug's bite.

among immobile folks—immobility not just from age and illness, but from language and cultural and financial and legal barriers as well.

When we considered these principles—comparing them to our centralized utilities or to conventional microgrid fixes—we realized that they all reduce the Bug's corruption in a new way. They're actually practicing "coherent integration."

Remember what we said about adding or interfering when we were talking about harmony? The trick is always this: You combine lots of different points of view, but the combining has to be "coherent"—it has to combine the views "in phase," so they add up instead of interfering with each other. That's why we engineers use things like "phased arrays" and interferometers: they let us see details no one detector can see alone. Stuff way out in the cosmos, or tiny things inside cells or atoms.

It's the same with things that go on in life, that don't make any sense either, until you listen to folks with different experiences, different viewpoints. It's a universal principle.

OK, you can see where I'm going with this. Our neighborhood mentors are working together on shared goals. They're committed to mutual benefit, to diversity, to inclusion, to connection, to truth. They're in phase.

And you know what? *We can join.* Together we have a chance to make the highest expression we can reach of intelligent life. In that reach we can "integrate." This is a way for any expressions of intelligence anywhere to extend to each other. In that stretch we can actually grow, we can find "more" of ourselves in others. So that's what I mean by "coherent integration."

The neighborhood organizers agree to be our mentors. We'll build new kinds of microgrids. They may look like a bunch of batteries and solar panels and green-fueled generators, but that's not what they are. If you focus on the package, you miss the essence. These are real microgrids. They have no borders. They're made of commitment and trust and extension, to and in each other. That's a good definition of love, I think, and of faith.

Coming home from Chinatown or Chelsea, we try to absorb our new lesson in connection. Is this work a way to "express our relatedness to a central order," in Heisenberg's words? Is it related to our search for "extraterrestrial intelligence"? For how quantum discoveries stretch

*Together we have
a chance to make
the highest expression
we can reach
of intelligent life.*

our view of reality? For how we can wrestle with the Bug in all its costumes?

Might this give us a lasting, defensible, logical moral compass?

A Moral Compass

Human civilization has depended on some weak strings to hold it together. Ancient religious commands. Fear of consequences, here or hereafter. An evolved empathy in our brains. But lately it seems we've been into more that divides us than unites us. In this era of distrust, we'll have to do better. We have to look farther. We have to find self-interest in other-interest.

If we're going to teach that to our grandchildren (or each other) and make it stick, it has to be based on new thinking. Beyond Sunday school lessons (even Unitarian ones). Beyond rules or revelation or pretty stories. There has to be good reason to believe it.

Remember the evidences of intelligence we said that we'd accept anywhere in the universe? The ability to recognize order, the freedom to choose it, the capacity to communicate with some purpose? That potential, that field of intelligence, is everywhere available.

The problem is, that potential can be used

for any purpose. Choosing order, communicating, and learning are evidences of intelligence, but they aren't always beneficial. Look at human history. Those basic capacities are not *in themselves* a moral guide. One of the first evidences of intelligence that we identified was the freedom to choose. Our compass must lie in *how* we choose.

Well, all the *isolated* forms that are familiar to us get chewed up by the Bug. What is not chewed up? Connecting realities that displace the Bug. The uses of intelligence that we found in our SETI conversation and in our microgrid neighborhoods. Healing, trust, learning and teaching truth, commitment and extension to each other, seeking beauty—these support what we called "coherent integration." They invite diversity, they cure isolation.

These ways of reaching to each other offer a way to be *more*. Self-interest is OK if it isn't limited to benefits that are owned exclusively. Many can only be shared. Mutual benefits come from commitments that really integrate. That's a way to be in "harmony" with the whole, with the central order. It's a way to be more. That can be lasting self-interest.

So, our logical moral compass points us toward connecting with each other truthfully and to our mutual benefit. Now we can be less fuzzy about what we mean by "benefit"—we mean helping each other's fuller and more truthful expression of these intelligence-capacities. Will my decisions and actions lead to greater freedom, organizing, communicating, building, learning, healing, loving, beauty, as we have found them? That can be our test of moral behavior. It gives us a moral compass that's rational—faithful to truth.

Of course, this is not really new. But now we can say it has a logical base. We've been studying the smallest things, the largest things, and the most critical things. The quantum world, intelligence in the cosmos, and life or death in our neighborhoods. They all point the same way. Self-interest is other-interest. We have a logical reason to believe in it, to live by it.

Well, how many of us really do live by it all the time? Not me, I confess. Can't speak for you, but I think not many. Most of us carry heavy regrets, remorse, shame. Shall we despair, then? Or is there some kind of logical grace, some rational forgiveness?

Healing,

trust,

learning and teaching truth,
commitment and extension
to each other,
seeking beauty —

— these support what we called "coherent integration."

They invite diversity, they cure isolation.

I can't go back and fix what I hurt or do what I didn't. But I can be faithful today. I can share my best with others, in the ways we've discussed. The central order of the universe is forgiving in that sense. I have not lost forever my ability to help, or to seek truth. Nor, thankfully, have those to whom I may have been unfaithful. They may still carry the pain, and I must still carry the burden of shame. I can't offer a way to put that burden down. What we've learned, however, may offer a kind of comfort and forgiveness to us all.

When we discussed "coherent integration," we considered how to be more than isolated individuals. What I give to others—my extension—becomes the more complete part of my identity. I can learn some fragment of truth and share it, or share some beauty, or heal others, or organize helpful things, or communicate truth more clearly, or set others freer. In this understanding of benefit, I can add to the whole expression of intelligent life.

Those whom I may have hurt can do the same. In that we share the same truth, we express the same enabling intelligence. Through them we connect with each other, we add to the "whole each other."

To me, a commitment to this extension suggests something like a logical prayer. At the start of each new day, I can acknowledge in remorse my unfaithfulness, I can lift my shame to carry it through the day. But I can also lift my gratitude, my acknowledgement in amazement of the opportunity to be faithful today. I can lift my commitment to truth and the benefit of others, my trust in this course, and my extension to the whole in which we all live.

And, as Mother Teresa counseled, we are called not to success but to faithfulness. I can then dare to lift my hope.

Mutual benefits come from commitments that really integrate.

That's a way to be in "harmony" with the whole, with the central order.

That can be lasting self-interest.

Hope

It rains, the whole world cries.

In times of loss or pain, we feel alone, and unloved. Not long ago, I lost a daughter and my spouse, my constant partner, my Polaris. I know the feeling of being lost, the world gone gray, the nevermore. In our dark hours, hope seems impossibly remote.

> ...O for the touch of a vanished hand,
> And the sound of a voice that is still![29]

We all go through our desert places. Every touch and sound eventually vanishes. The whole world cries. Is love gone, then? Is all hope gone?

For me, they were brought back within reach by sharing our work with our neighborhood colleagues and by the search for intelligence in the universe. They're related.

Why should this work help? It's not just

---

29. Alfred, Lord Tennyson, "Break, Break, Break," in *Poems*, (London: Edward Moxon, 1842).

a distraction. Is it real? Logical? It does not restore the touch or the voice. Their isolated, aging source is gone from this troubled world.

But it seems a small notion of life that could be so easily cut off. Looking back at the great sweep of a universe lighted by a central order—at the discoveries of quantum realities way beyond our experience—at the "coherent integration" of commitments in our poor neighborhoods—one cannot escape the sense that life doesn't stop at the edges of our brains or our Flatland.

Oh, of course that Bug chews up our brains and our bodies. Of course, we can explain the origin of all our molecules out of stardust. We can make artificial intelligence, and manipulate our own brains. Sometimes it seems we have a smug explanation for everything, and it will all end up in nothing, as it began from nothing.

Which of these views of life shall we accept? I think we can agree that both are valid. We just have to admit there can be *more* to reality. That means there can be more to "us," and more to relation. As Stephen Hawking said, "What is it that breathes fire into the equations and makes a universe for them to govern?...science... cannot answer the question: Why does the universe bother to exist? I don't know the answer to that."[30] We have to look farther.

Einstein has been quoted as saying that:

> A human being is part of the whole... He experiences himself, his thoughts and feelings as something separated from the rest—a kind of optical delusion of his consciousness. This delusion is a kind of prison for us, restricting us to our personal desires and to affection for a few persons nearest us. Our task must be to free ourselves from this prison by widening the whole circle of compassion to embrace all living creatures and the whole nature of its beauty.[31]

---

30. From "The Origin of the Universe," in *Black Holes and Baby Universes* (New York: Bantam Books, 1993).

31. Quoted in Heinz R. Pagels, *Perfect Symmetry: The Search for the Beginning of Time* (Simon and Schuster, 1985), 362. This does not suggest that Einstein believed in some kind of personal God, which his writings make clear he did not (despite his frequent use of the word).

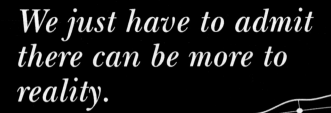

*We just have to admit
there can be more to
reality.*

*That means there can
be more to "us," and
more to relation.*

Erwin Schrödinger, the famous scientist who put quantum mechanics into a "wave" form,[32] offered a similar view of this "whole." He proposed that intelligence is a singular reality, that what we experience as separate forms of consciousness is a part of that reality:

> I should say: the over-all number of minds is just one. I venture to call it indestructible since it has a peculiar time-table, namely mind is always now. There is really no before and after for mind...we may, or so I believe, assert that physical theory in its present stage strongly suggests the indestructibility of Mind by Time.[33]

And Roger Penrose, mathematician and frequent collaborator with Stephen Hawking,[34]

---

32. Heisenberg used "matrix" mathematics to predict the behavior of quantum systems, and Schrödinger developed a "wave" model. They were later shown to be equivalent. Schrödinger also won a Nobel Prize, with Paul Dirac, "for the discovery of new productive forms of atomic theory."

33. Erwin Schrödinger, *What is Life?* (Cambridge University Press, 1967).

34. In 2020, Penrose was awarded one half of the Nobel Prize in Physics for the discovery that black hole formation is a robust prediction of the general theory of relativity.

offers a mathematician's view of our access to truth:

> ...mathematical truth is absolute, external, and eternal, and not based on man-made criteria...mathematical objects have a timeless existence of their own, not dependent on human society nor on particular physical objects...When mathematicians communicate, this is made possible by each one having a *direct route to truth*, the consciousness of each being in a position to perceive mathematical truths directly...communication is possible because each is directly in contact with the same externally existing Platonic world! [italics and punctuation Penrose's][35]

These instincts for a whole, a universal mind, an eternal truth, are of course consistent with Heisenberg's "central order." From different viewpoints, each is seeking the same reality. And so are we. That fundamental reality makes possible our integration.

---

35. Roger Penrose, *The Emperor's New Mind: Concerning Computers, Minds and the Laws of Physics*, (New York: Penguin Books, 1991).

Our task must be to free ourselves from this prison by widening the whole circle of compassion...

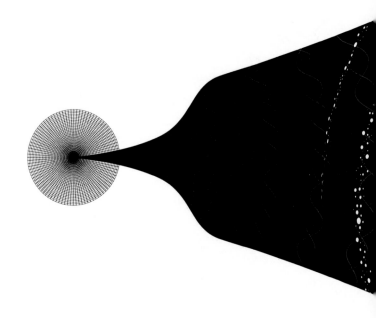

...to embrace all living creatures and the whole nature of its beauty. –

ALBERT EINSTEIN

Of course, our grasp of this is limited.[36] But our reach for it need not be cut short.

What is it that can integrate into that whole reality, that can express intelligence or mind, that can be faithful to truth? We can make commitments to truth and each other that "integrate coherently." We all make such commitments when we're at our best. So, no doubt, do countless others around an infinite multiverse. All have a "direct route to truth." All express the same universal order and intelligence. All such faithful expressions of intelligence, whatever their local mechanisms, make the same reach toward the whole reality they share.

And, as Schrödinger said, that whole is timeless. That's hard for us to understand, but it's perfectly logical. Our time was born with our universe. It started some fourteen billion years ago. And its direction is given by—guess what?—that same old Bug. The Bug's third costume is

---

36. Kurt Gödel's famous Incompleteness Theorems showed our inability to reach absolute proof of even the most basic premises, because we're inside the puzzle we're trying to solve. ("Provability is a weaker notion than truth," to quote Douglas Hofstadter's paraphrasing of Gödel's theorems.) See Douglas Hofstadter, *Gödel, Escher, Bach: an Eternal Golden Braid* (Random House, 1989), pp.15–21.

time. Without the Bug, time could run as well backwards as forwards. That is an established fact. As Hawking points out, our psychological sense of time is due to that same Second Law of Thermodynamics—our experience of time and the cosmological direction of time have the same "arrow" as this thermodynamic direction.[37]

So, let's try to adjust our focus. Einstein's diagnosis—our near-sightedness—obscures much of reality. It comes from evolution and conditioning, of course, on our Flatland. Understandable, perhaps excusable. But that thin slice of our experience—that local result of evolution—that container—just isn't the whole story. There's more to it, I think we can agree.

Of course, it's hard to picture this "more" of ourselves. Maybe my friend the electron can help. I've probably used more electrons than any of you. Over years of energy and radio work, I've developed a special affection for them. Like us, the thing about electrons is they are what they express.

Leon Lederman says he thinks of Lewis Carroll's Cheshire Cat. (Carroll himself was a

---

37. Stephen Hawking, *A Brief History of Time* (New York: Bantam Books, 1988).

mathematician, and his allegory had substance as well as bite.) Thus, says Lederman:

> ...the electron is real. Probably a point particle, but with all other properties intact. Mass, yes. Charge, yes. Spin, yes. Radius, no. Think of Lewis Carroll's Cheshire Cat. Slowly the Cheshire Cat disappears until all that's left is its smile. No cat, just smile. Imagine the radius of a spinning glob of charge slowly shrinking until it disappears, leaving intact its spin, charge, mass, and smile.[38]

George Johnson, who writes about science for *The New York Times*, considers particles like the electron to be "all information":

> A marble *has* mass, color, size. But an electron *is* mass, spin, and charge. A particle is completely defined by its quantum numbers. It is all information. Spin ½ plus 1 unit of negative charge ($1.6021892 \times 10^{-19}$ coulombs) plus a mass of 9.1×10^{-28} grams

---

38. Leon Lederman and Dick Teresi, *The God Particle: If the Universe Is the Answer, What Is the Question?* (New York: Dell Publishing, 1993). Yes, I recognize that string theory suggests that concepts of point-sized particles can be considered vibrating loops of "strings" that have finite extent, perhaps in ten space dimensions.

is an electron. These are not just labels or qualities exhibited by something underneath. There is nothing underneath.[39]

Electrons are real. They're not a special case, not something outside our everyday world so we can ignore them. Like the Higgs field and the central order in the universe, we can't see them with our Flatland senses but they make us possible.

In just the same way, what we really care about in each other is not what's "underneath." The Bug chews that away. I propose that what we care about in each other (and ourselves) is how we communicate and teach and learn, how we free and heal each other, what we build, what beauty we see and make, our commitments and trust and extension to each other—all those real evidences of intelligence that we identified back at the SETI telescope in Harvard. It's how we express the same field of intelligence, how we share the same truth.

We experience disintegration and time because we live with the Bug in our world. The Bug limits the means available to us for

---

39. George Johnson, *Fire in the Mind: Science, Faith and the Search for Order* (New York: Vintage Books, 1996).

expressing intelligence. The same must be true for thinking creatures everywhere, who likely evolve very different means. Fortunately, the means are not what we hope to integrate. Our hope rests on the nature of a central order, making possible our coherent integration in a whole reality. To that reality we can reach, whatever our limited means. Our reach is never perfect, but it can be faithful.

This concept of a timeless "whole" isn't mystical or other-worldly. It just seems distant from the packages we walk around in. We got where we are by a billion years of competitive survival-of-the-fittest. We evolved to breed and defend a "self" in a hostile world. The amazing bodies and brains we've inherited are the result. They haven't had to look for "more," so they're not equipped to see it.

And for most of human history, we've put our hope in the wrong image—in an individual, isolated "soul," somehow freed from the body when it dies.[40] It is comforting for many, espe-

---

40. Attempts to detect and even weigh the departing soul have been made for centuries, even capturing wide media attention as recently as 1907, when a Massachusetts physician named Duncan MacDougall set up an elaborate scale system and reported a finding of 21 grams.

cially in loss. It's familiar, it's the way we experience the world.

But it doesn't lead anywhere. Every isolated package is subject to decay. Without a sustained, purposeful input of energy, the Bug chews it up. So, we lose the forms, the packages in which we sense love and the other expressions of intelligence. They disintegrate. Like the blossom, we can't restore them. Those packages are subject to the Bug's chewing away, and of course they change all the time.

So, the touch and the sound that we and Tennyson miss are gone—lost to time, to disintegration. We can't change reality. But we can change how we experience it. In particular, we can experience more of it.

Of course, one can dismiss all things that can't be measured. To me, that is no more logical than dismissing everything not "revealed" to the faithful in a particular religion. Look:

Quantum and relativistic physics have shown us how much more there is to reality, that reality including us. At first, we tried to square the discoveries with our human experience. But reality stretches far beyond such excuses. It isn't the quantum world that's weird; it's our Flatland world-view that's incomplete.

For most of human history, we've put our hope in the wrong image—

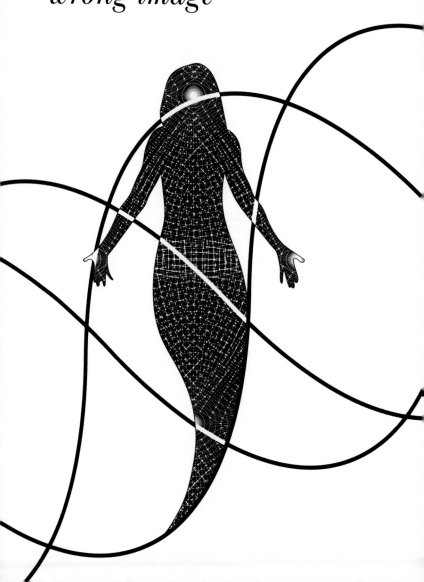

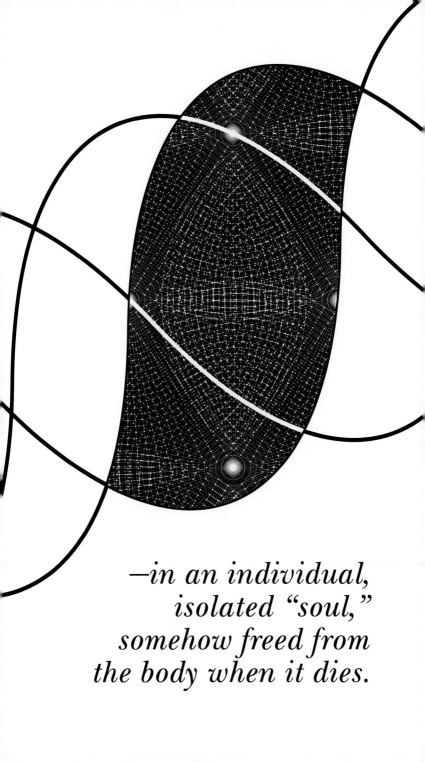

—in an individual, isolated "soul," somehow freed from the body when it dies.

The search for extraterrestrial intelligence has shown us what would qualify as intelligence, and how to relate it (and us) to a central order in the universe. At first, we tried to define intelligent life as organic, carbon-based, water-dependent biology. But the reality of intelligence stretches far beyond such conceits.

Reducing the Bug's bite among committed colleagues in vulnerable neighborhoods has shown us an application of these realities that's within our reach. At first, we tried to apply our solutions to our view of the problem. But the mutual extension that we found there offers a way to stretch who we are.

Once we free ourselves from the myopia that Einstein was talking about, from that Flatland conditioning, from that sense of being wholly contained in our packages, we can consider our wider being. In every form of pain, our hope is to escape that isolation, to extend our experience, and to be loved.

The touch and the sound of those whose love I miss were chewed up by the Bug—by disintegration, by time. So, in time, will mine. But what we experienced in that love did not depend on the sound, the touch, the sight. Those could be reproduced. They changed all the time. It was really the experience of commitment (knowing it's unconditional), trust (being able to count on it, to be vulnerable), and extension to each other (not being isolated, becoming more together). All that, of course, is an expression of intelligence.

Can we feel loved by a universal intelligence, a central order of the universe? It seems a rather cold embrace. Remembering what we really mean by love, however, our answer can be Yes: that field of intelligence, that whole each other, offers a perfect commitment of truth and order, no matter our choices or condition—a perfect trust in our ability to choose faithfully, no matter our past—and a perfect extension, in its inclusion of all faithful expressions, no matter how inconstant our expressions may be.

But can we accept such a perfect offer, and feel loved by it? Back in our imagined dialogue with the SETI scientist, we agreed that love should add to each other, to our fuller expression of the qualities of intelligent life. Can we add to an already-infinite intelligence?

Well, we can take on the nature of the whole, inclusive intelligence as best we can understand it. In putting useful things together, in our good work, our learning and teaching and

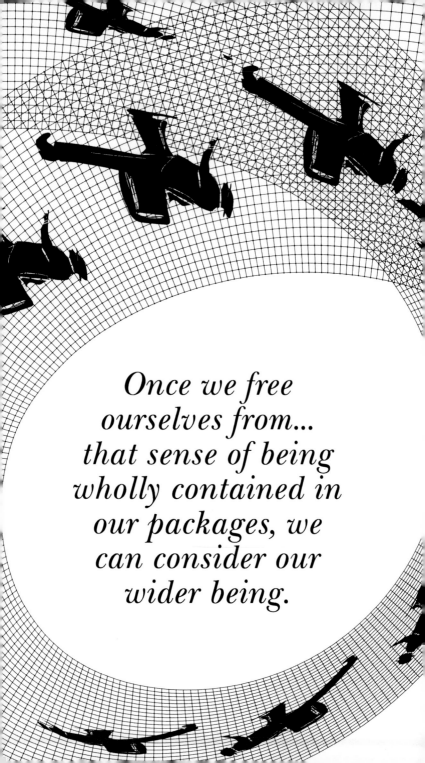

*Once we free
ourselves from...
that sense of being
wholly contained in
our packages, we
can consider our
wider being.*

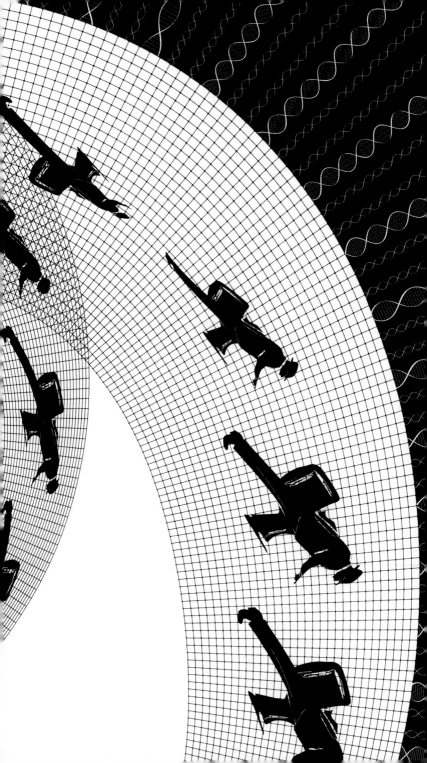

communicating truth, in healing and loving and making beauty, we can extend ourselves to others. We've learned something about that "coherent integration." Indeed, it can add to the more complete expression of intelligence.

That extension, with our commitment and trust, would be a rational faith, our love of the whole. We can then feel loved in return—not as isolated souls but as integrating expressions, adding to the whole in which we live. By that addition we're included.

Through our truthful commitment, our beneficial extension to those we can reach, we can express our love for the whole in which— one might say in whom—we live. Those whom I love make the same addition, by the same faithful extension to me and others. As I work with those making similar commitments to truth and each other, I feel increasingly connected with all. The more faithfully we add, the more coherent our integration, the more I can actually experience that inclusion.

This is not to suggest some kind of personal God, caring in mysterious ways for favored creatures. Nor does it suggest an isolated, immortal soul inside our mortal brains. Reality is better than that. We can choose to be part of the whole intelligence that makes the multiverse.

In that choice of truth and benefit, we can join others who do the same. Our "coherent integration" together then adds to the whole. We can see that faithful expression of intelligence as the "more" of us. Such an expression can join the "more" of others, including those we've loved. It's free of disintegration and ignorance and time. That joining, that stretch to the whole each other, is within reach. It only seems remote because of the limits of our evolution.

So, I can be faithful to those I love—whether physically present or not—by extending to my colleagues and our shared work. For me, this is healing. I can really experience being part of a larger reality. We're joined by those same practices and commitments that we loved in each other. Those I have loved have integrated coherently in the same whole.

That's where I find those I love, seen or unseen. In the whole, timeless reality of intelligent life, which is made up of all our faithful expressions. If I listen for it, I can hear their love. That voice is not still, after all.

When I go to work tomorrow morning, I will find those I loved. Those whose touch is vanished from the earth, whose voice is still, are yet present and entangled, in our good work together.

Life doesn't stop at the edges of Flatland. In this expanding view—of the universe and our place in it—we can find hope.

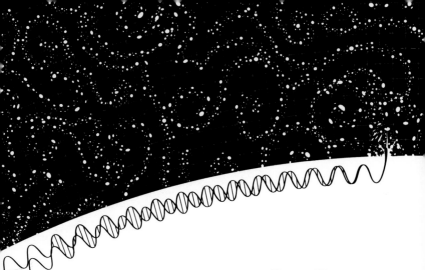

*So, I can
be faithful to
those I love—*

*whether physically
present or not—*

*by my commitment and
trust and extension
to my colleagues and
our shared work.*

For me, this is healing.

About the Author

David Sayre Dayton now works fulltime developing "microgrids" for emergency power in low-income neighborhoods threatened by increasing storms and flooding. He has been introduced as the "father of our industry" at national conferences, patented and published energy and communication techniques that anticipated today's cell phone and energy-efficiency industries.

The sciences behind all these industries displace what we call "entropy." Sayre shows how entropy also measures our experience of time, ignorance, and death. In this book, he links the work of some of our greatest scientists to that of his colleagues in vulnerable neighborhoods. Together they support a wider, hopeful view of intelligent life.

Around these same sciences, Sayre has formed a dozen companies, profit and non-profit. These have led him to encounters and places of great human striving—in prisons and mental institutions, laboratories and cathedrals, the Pentagon and the White House, neighborhoods of the very rich and the very poor. He derives from these encounters and sciences, and from his personal loss, a rational moral compass and a way to comfort.

Sayre is an engineer and entrepreneur, the father of five children and four books: *The Great Improbability* (Peter E. Randall Publisher, 2010); *Something There Is* (Peter E. Randall Publisher, 2014); *Flatland*, (with award-winning children's illustrator Rebecca Emberley, Two Little Birds Books, 2014), recommended by Parents' Choice Awards; and *The Flatland Dialogues* (Peter E. Randall Publisher, 2017).

About the Illustrator

Evan Robertson is a self-taught illustrator living in Bucks County, Pennsylvania, who focuses on bringing great language to life. As a student of classic literature at Yale and of drama at The Juilliard School, he is inspired by beautiful language that has stood the test of time. He has partnered with the Library of Congress, the Museum of Modern Art, Penguin Random House, Doubleday, Abrams Books, Barnes & Noble, and Indigo Books. His work is available through his creative studio, Obvious State (obviousstate.com).

About Our Cover

Those who like such things may have noticed a hyperbola on our cover. Yes, and an ellipse partly superimposed on it.

We chose an upward-facing hyperbola to represent all of reality, all of truth, which is infinite. The ellipse represents the experience or perception of intelligent beings anywhere in the multiverse. Part of that is within the hyperbola and thus is truthful, part lies outside the hyperbola and thus represents illusion or falsehood or deception. The part within is open to integration with others, making up the whole of intelligent life.

And the base? The grassy base resembles a graph of all the frequencies in "white noise," which theoretically contains infinite information. If you could slide two samples of white noise against each other until they lined up perfectly, that "autocorrelation" would be an infinite spike, a singularity—represented by the "stem." The whole thing looks rather like a tulip, whose blossom represents the beauty of shared truth.